EFFICIENT PROJECT MANAGER

EFFICIENT PROJECT MANAGER

A PRACTICAL GUIDE

Fawzi Belguet

gatekeeper press™

Tampa, Florida

Efficient Project Managemer: A Practical Guide

Published by Gatekeeper Press
7853 Gunn Hwy., Suite 209
Tampa, FL 33626
www.GatekeeperPress.com

Library of Congress Control Number:

ISBN (paperback): 9781662944987
eISBN: 9781662944994

To my late Mother Fatma Zohra;
Who gave me the gift of unconditional love.
I miss you, Mom!

Table of Contents

Chapter 1:
Introduction

What is a Project?

"Real life is, to most men, a long second-best, a perpetual compromise between the ideal and the possible"

Bertrand Russel

Chapter 1: Introduction

"Bathroom Renovation Projects!"

We spend most or a big part of our life planning and executing small or big projects instinctively without even thinking or noticing that we are "Project Managers". Some call it "Common sense" or just a "normal, and logical" series of sequences that leads to achieving our goals. A lot of learning material that delve into the subject matter of project management and in a detailed, quantifiable metric is available. A lot of schools, university classes, and certifications (like PMP) offer an extensive body of knowledge that support future "PMs" to be Effective Project Managers.

This book is not about adding another academic document to the already well-established scholastic Project Management library, but more as a practical guideline that I wish I had when I morphed from an Engineer to a Field Engineer to ultimately be a Project Manager. Throughout my 20-year career in the field, I have noticed that corporations, consultants, and construction companies lack documents (guidelines) or a set of steps that specifically address the handling of projects based on each project application. A project manager should be able to use his/her skills to tailor and custom fit his/her approach based on the project specifics, i.e. The same project with the same general scope could drastically change based on the geographical location or even political status and a lot of other aspects to be considered. A new project manager in a pharmaceutical company does not have and should not have the same priorities or use the same approach as a pipeline or utility distribution project manager!

So many "PMs" take, if not a couple of years, at least a few months to sort it out and become efficient. In spite of all the right certifications and degrees, I have witnessed many PMs (and sometimes the entire department) struggle, going through this painful road of learning, navigating through the abstract road of managing a project or many without a consistent, balanced approach.

This book is an attempt to lighten the road and show a simple, clear path to follow to be successful. I've started this book with a few simple examples or scenarios that ease the reader into the project manager world, but also to –by using common sense–show that each one of us is a potential, successful PM as long as he/she follows rational, standard steps that work and apply to all projects at the same time, be aware of each project constraints.

Then, we merge into the organic structure of a project in generic details. The word "Generic" means to be customized based on each project discipline, industry, and application.

I refer to a project as a story that needs to be told and shown, and the execution of the project is nothing but a process that needs to be put together to faithfully showcase the story to the viewers. Let's set the stage with these two scenarios:

Scenario 1:

You bought your dream house and settled in with your spouse and children enjoying everything about it except that master bathroom with outdated vanity, not-too-efficient lighting, and a fan that cannot keep up with family activities, not to mention the paint. So, with your spouse, you've decided this "tax" return year this project must be complete!

Renovating the bathroom this tax year is the most important decision you will ever make. Around the dinner table, you sit down and start drafting your "Wish list" based on your needs and desires, looking around for the right material and options, virtually and in person, refining and finalizing your list, leaving some open items for later on. Typically, your tax return is around $5,000, but in order to renovate your dream bathroom you you'll need at least $15,000. The difference must come from your savings accounts with the assumption that the cost of labor is included in your budget. At this stage, the timing is not as important as the $15,000 limit that you cannot, and do not want to go above, you emphasize the budget!

Scenario 2:

Your spouse called you to let you know that the master bathroom toilet is leaking. Your first reflex is to make sure to "Limit" the damage and avoid catastrophic outcomes, then look online, make phone calls, and schedule an emergency plumber to come and fix the problem. At this stage, due to the urgency, you do not negotiate the rate, but you do emphasize the urgency (time) as not negotiable!

Scenario 1 & 2 both are projects that require scope, budget, and time. Both are about the same bathroom in the same house and the same owner. However, the difference is the constraints that apply to each case.

Scenario 1 biggest constraint is the budget, or how much money we are willing to spend to meet our goal.

Scenario 2 is an emergency that does not allow you the luxury to think about material or rate, but you know that time is your paramount priority "Get it fixed" at any cost!

So…

What is a Project?

Who is the Project Manager?

A Project:

A set or series of organized, interconnected actions that lead to the development of an idea (a goal) from thought to completion.

Any set of actions that requires planning (Scope), money (Budget) and time (Schedule) is A PROJECT!

It is ultimately an improvement process!

A Project Manager:

Most of our life is nothing but a series of projects (small or big), occurring either:

1. In series (back-to-back).
2. Overlapping.

3. Simultaneously.

4. Recurrent (repetitive).

5. Or a combination of some or all of the above.

All of us, at least once in our lifetime, had a thought or an idea, set a goal, explored it, planned it, budgeted for it, and finally realized it. We either oversaw or delegated tasks and followed the progress from inception to execution. These goals (ideas) could be as simple as going to work, to the gym, or organizing a barbecue.

Thus…each one of us is a potential Project Manager!

What is the foundation of a project?

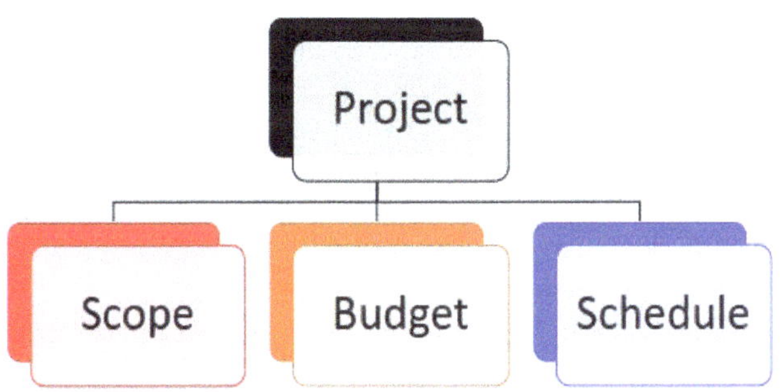

Project Foundation Tripods:

To take an idea or a goal from a thought to a reality you must build an infrastructure that can support it and lay a good solid foundation. Thus, any project must be built on three main, interconnected components or elements: "Scope," "Budget" & "Schedule".

These tripod structures are the genesis of any project and its success is completely dependent on the accuracy and risk assessment of each component.

The next chapters will explore each one of them in relative depth, and allow you (the project manager) to successfully achieve any project's objectives.

Chapter 2:
The Scope

Telling the story (The book)!

"An army of sheep led by a lion is better than an army
of lions led by a sheep."

Alexander The Great

Chapter 2:

"The Scope," "The Book," "The Story":

The scope is the most important document in any project's success. It is ultimately the roadmap to achieve the idea's goals. The scope must tell "The story" in detail, from visualization to performance. So…

What should a scope include?

Back to the bathroom project…

1. **The idea:** Renovate the bathroom! This idea is simple as it seems to be the seed to a process that is called "A Project". The action or process of drafting steps to achieve it is called "The Scope".

The scope is a process of capturing a series of comprehensive, logical actions in an organized way, within the industry or field-specific applied codes, standards, norms, or laws (Local, State, Regional, Federal, or International) to achieve a set of "Well Defined" goals and performance objectives.

2. **Project's Goals:** Replace bathtub with standing shower, install floor radiant heat, and more.

If a project is a tree, then the idea is the seed and the goals are the fruits of that seed. If the growth process (project) is not well maintained, nourished, fertilized, and properly watered, then the fruits will either be bitter or nonexistent.

Project goals vary. They come in different forms and shapes, completely depending on the needs, disciplines, and situations. They could be artistic, healthy living, or any other expressions, or shapes, they could be as simple as having a picnic and as complex as nation-building.

Setting the right goals or objectives that transform an idea into an achievable, realistic entity is the key to any successful project. Below are some examples of standard objectives or goals that the project can carry:

a.	**Safety, Security (Cyber or In Field):** One of the biggest issues with safety is that it cannot be quantified, simply because if you allocate in the budget a certain amount to establish a safe environment and no incident occurs, the project just goes through its natural course with no interruption. On the other hand, if you underestimate or minimize safety allocation in the budget, taking shortcuts, then a lack of commitment to safety leads to incident(s) with tangible, quantifiable, and most of the time very costly consequences. Thankfully, safe practices became part of industry standards, local, federal, and international, and must be considered in each project's scope.

To Remember:

"You cannot put a cost to safety, but the lack of it could be very costly" F.M. Belguet.

"Cheap; is most of the time the most expensive option" F.M. Belguet

b. Business Goals:
   i.   Performance.
   ii.   Expansion.
   iii. Upgrade.
   iv.   O&M Improvement.

3. The Feasibility Studies (Could it be done?): Would the bathtub fit in this space?

The main concern that faces any project manager is transforming an idea into a functional entity, regardless of driplines. The first critical aspect of this transformation is answering the question: Could it be done?

Answering this simple question could be as simple as taking a tape measure or as complex as being a project in itself with years of research. The answer has also to take into account and consider the Space/ Time/ Budget (STB) constraints.

Is it feasible within the allocated or allowable Space/Time/Budget?

Once this answer is determined then we can start framing the project's visions (picture) by addressing the following questions in detail:

c. **Engineering:** How?
d. **Budget**: At What cost?
e. **Schedule:** How long?
f. **Permits:** Authorized to do (Make, Build, Use) it?
g. **Codes:** What are the applicable industry standards and codes?

Define Owners/Scope Reviewers/Stakeholders:
The House of Representatives
Operations
Engineering
Environmental
Legal
Business Development
Safety

4. The Stakeholders: If a scope of work is the story, then stakeholders are the main characters!

Three phases:

h.   Project Stakeholders at inception: Most of the time the entity/ies that proposes the idea, as business developer, owner, and any direct beneficiary (impacted) of the project. **Initiation Team.**

i.   Project Stakeholders at realization: The team responsible for making or bringing the idea to reality, like engineers, cost controllers, and contractors. **Design/Construction Team.**

j.   Project Stakeholders at reception: Directly impacted by or users of the project, Operations, Sales, and more. **Commissioning Team.**

5. **Scope development process. "Tell the story":** After preparing the right pot and soil, now the time to plant your seeds has come and you are ready to start the process of writing the scope in the right conditions for the project to be successful.

Any successful scope must be developed to meet the below minimum criteria:

1. A document or script that meets the owner's expectations.

2. A design to pass the Performance Acceptance Test PAT set by the client.

3. Feasible and Realistic.

4. Meet the industry or related field standards.

5. Well defined by:

    i. Objectives.

    ii. Performances.

    iii. Location(s).

    iv. Permit requirements.

    v. Standards, codes, and regulations.

    vi. Bounded by: Space (Real or Virtual), Budget, Schedule, and Timing.

Scope Process Flow Chart

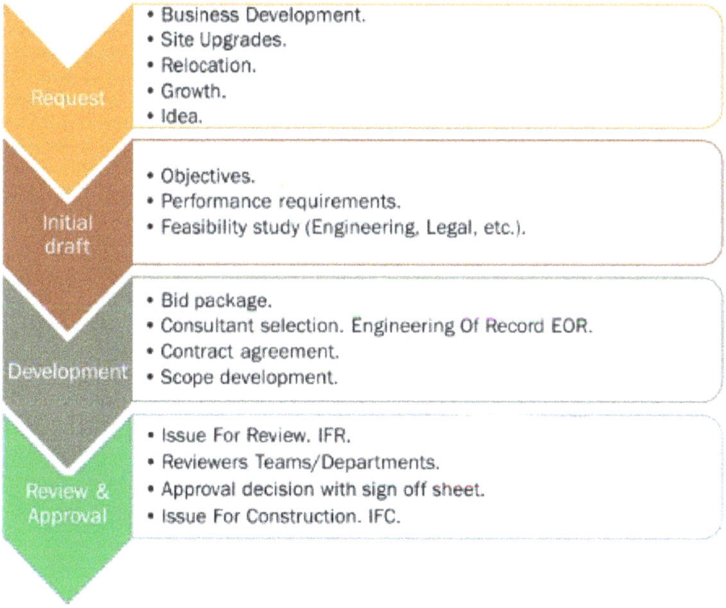

Request
- Business Development.
- Site Upgrades.
- Relocation.
- Growth.
- Idea.

Initial draft
- Objectives.
- Performance requirements.
- Feasibility study (Engineering, Legal, etc.).

Development
- Bid package.
- Consultant selection. Engineering Of Record EOR.
- Contract agreement.
- Scope development.

Review & Approval
- Issue For Review. IFR.
- Reviewers Teams/Departments.
- Approval decision with sign off sheet.
- Issue For Construction. IFC.

Chapter 3:
The Schedule

Movie scripts!

"Timing is everything Warrior"
Sun Tzu's - The Art Of War

Chapter 3:

The Schedule - "The movie script"

A project schedule is a timely script that rationally ties specific actions in a certain order and logic, leading to achieving project goals that transpire through and from the scope.

However, in order for this transcript to become "playable" it must follow a certain criterion, taking into consideration the complexity, the timing, and the location (s) of the project (Movie scenes).

1. **Complexity:** The degree of complexity can be a deterrent to the realization of any project. For the same areal distance between two points A and B, the path could be a straight flat highway or could be unsurmountable mountain hills and water crossing surfaces. Even though the goal is connecting A to B, however, the complexity of the second path makes it impossible to achieve, at least within the available time frame and budget.

2. **Timing:** There is a distinction between the schedule and timing. Timing the construction could make or break the project the same way the path complexity does. Building a steel storage tank in a desert in the summer is very problematic from a safety and

production standpoint. The execution of the same scope, pre-COVID-19 vs during COVID-19, will not take the same time due to added safety measures and other delays caused by the pandemic. Myspace and Facebook had the same goals, targeted the same customers, and a lot of other similarities, but one of the key factors that made one more successful than the other is again, timing. At inception, the successful platform had all technological tools and means aligned and perfected to meet the project goals. That was a good time to bring this project forward and the rest is history!

3. **Location (s):** "Location, Location, Location." In today's era, location here applies to real and virtual places where the project success depends (see above analogy Myspace vs Facebook).

A schedule cannot compile and combine a series of actions that lead to nowhere! All projects need location, paint, and a canvas, just as a house needs a space for its foundation. A pipeline or program needs to start somewhere and terminated somewhere else All of them need a beginning and an end. Thus, **a space must be secured for the project!**

4. **Schedule Flow:**

In order for a schedule to be as efficient and as realistic as possible, it must follow a certain rational "Natural growth" based on the facts and needs (goals). Any scheduler or project manager must acknowledge the following points:

1. **Project constraints:** Depending on the nature of the field and scope, the constraints should be known, determined, exposed, and captured in the schedule as "Milestones" and as early as possible. There are three constraints:

i. **Time:** Scope development, procurement or constructability obey certain laws of distribution that sometimes cannot be bypassed. The chain of supply for instance is still suffering from COVID-19, and what it took eight weeks to get pre COVID went up to 16 weeks post-COVID.

ii. **Space:** From construction to hauling site(s), the project location is one of the key factors to build a realistic, good schedule, but unfortunately, often overlooked especially at the beginning of the project (construction). A space (s) that allow equipment and manpower to safely operate should be one of the constraints and conditions for a successful project. Material and equipment hauling is also a factor to not be underestimated. An onsite storage area, multilocation sites, or 50-mile warehouse are to be looked at from the logistics standpoint and, thus, schedule.

iii. **Permits:** Major to minor permits, local to federal, and right of ways to site exploitation are the keys to opening and closing a project. Thus, they must be (depending on the complexity of the project) considered and accounted for in the schedule milestones as well.

2. **Build it!** Once the project constraints are determined and well-defined, then a detailed schedule can be built. Typically, a schedule is a timely script that contains standard generic sequences as illustrated in the chart below. It is certainly not the only applicable script, depending on industry applications and disciplines, that chart could be modified to suit any project's specifics.

Schedule: Project Timely Script

This is the foundation of any project schedule. Naturally, depending on the complexity or specificities, the block could be altered to a certain extent to meet the goals. However, most of the project's lifecycle goes through a similar growth flow.

3. **Contingencies:** It is a time margin of error (s) for a project, or task-specific predictable or unpredictable circumstances to counter potential delay (outlier).

   Total contingency could be:

   i. **Cumulative/Dependent:** Or a total added time from dependent-linked tasks. For example, equipment installation is completely dependent on the equipment receipt time (or delay), which in itself will dictate the installation timing. A four-week delay could shift the installation time to harsh weather conditions (cold or hot) and be a factor on an extended or deferred installation schedule, which in itself will impact commissioning, etc.

ii. **Overlapping/Independent:** When two independent tasks A & B are subjected to their own respective contingencies, a project manager or a scheduler has to consider the following timing factors: (A & B refer to contingency time)

   a. Starting at the same time: If A > B, then use A contingency.

   b. Finishing at the same time: If A > B, then use A contingency.

   c. If A finishes before B: Then use B's contingency end date.

iii. **Combination:** At the end, your total contingency should group both cumulative and independent contingencies based on task dependency links. It is very crucial to note that over-estimating project schedule contingency factors could have a significant impact on the capital, budget, commissioning, and equipment warranties among other negative repercussions. The assessment should focus on the predictable, factual risks and not a probability analysis.

4. **"Hold" points:**

Any project schedule—depending on the complexity and disciplines—must include and quantify "Hold" points. It is a series of key transitional phases and evaluations that allow to:

1. Assess the project's progress (Schedule and Budget).
2. Identify schedule deviations and rectify or reduce incurred errors.
3. Evaluate deviation consequences on the remaining schedule.
4. Revised and updated the schedule as deemed necessary.
5. Revised and updated the budget as deemed necessary.

One of the most commonly used "Hold" stops are engineering review times, set to be in general at, initiation (0%), review at 30%, 60%, 90%, and IFC. But each field or industry has its own metric and references when it comes to identifying "Hold" stops and determining their respective duration, but an accurate, realistic schedule must include such "Review".

Chapter 4:
The Budget

Movie production!

"Too many people spend money they haven't earned, to buy things they don't want, to impress people that they don't like."

Will Rogers

Chapter 4:

The Budget - "The movie production"

Any action, task, service, or development requires time and effort, thus a budget to financially support the necessary actions to bring the project to its fruition and realization is needed.

The accuracy of the budget is completely dependent and linked to the phase of the project. In the same way the schedule is developed and used, the budget is subjected to contingencies, constant re-evaluation, and revisions depending on past, current, and future expectations. For example, budget evaluation or estimates of pre and post COVID-19 are completely different, the chain of supplies has been drastically impacted by long lead time deliveries and inflation, and healthy site accommodation, just to name these three factors added a significant burden on any project budget, that was not accounted for pre-COVID-19. Thus, an accurate forecast must rely on and factor in the impact of some important currents, but also future events, like geopolitical, environmental and health conditions, and local and global currents.

Ultimately, a budget is an approximate estimate of a project, cost, time, and effort needed to achieve the scope goals. The accuracy is refined through the project evaluation process based on the accuracy of the present data. Below are some basic, typical budget estimates:

1. **Rough Order of Magnitude ROM (+/- 30% to 75%):** This is the initial evaluation at the beginning of the business development (or just "an idea!") level to have an order of magnitude estimate that determines if the project is worth pursuing or not.

2. **Accurate Order of Magnitude AOM (+/- 15% to 25%):** Typically, an estimate merges the feasibility study and scope development estimate, drawing a more accurate picture and allowing a refined evaluation of the project requirements.

3. **Final Order of Magnitude FOM (+/- 10% to 15%):** An estimate that identifies all the project ingredients, from material and equipment cost to workman hours. A project forecast basis that must be maintained and updated through field progress and project schedule evolution.

There are other estimation techniques, but for the sake of making this publication simple and easy to follow, the list was restricted to those mentioned above.

The below chart is a list of interconnected tasks or phases that support the project estimator to take them into account while estimating the budget.

The chart is a helpful guideline than a checklist. Ultimately, an estimation completely depends on the nature of the project and the applicable industry.

Project Budget Applications:

- Scope development/Engineering
- Consultation Fee: Legal & Facilitators
- Permits
- Site development
- Environmental
- Safety Accommodations
- Communication
- Material & Equipment
- Construction/Installation
- Commissioning & PAT
- In Service
- Internal Services: Ops & Eng.
- Project Closing: Docs & As-built

* PAT: Performance Acceptance Test.

Chapter 5: Communication "The Glue" Macro vs Micro Management Style (When? How?)

Movie Director!

"The mere physical man is like the ant crawling on the paper, who observes black lettering and attributes its production to the pen and nothing more."

Al-Ghazali

Chapter 5:

Communication – "The Glue"

In any project, communication is the glue that ties and links all project phases. If that link is broken or misused, regardless of a well-scope development, schedule accuracy, or budget precision, the project cannot be successful or at least reach all its goals.

> **"NASA lost its $125-million Mars Climate Orbiter because spacecraft engineers failed to convert from English to metric measurements when exchanging vital data before the craft was launched, space agency officials said Thursday."**

Los Angeles Times, October 1, 1999. [1]

The above unfortunate real-life event of losing the $125 million Mars Climate Orbiter in 1999 by NASA was 100% a communication failure and a reminder to all of us how sensitive, crucial, and essential to establish a comprehensive communication protocol to avoid such failure (or worse!) and guarantee project success. It is worth emphasizing that the loss was not just monetary: many years of efforts, research, coordination, planning, hopes and aspirations were also lost,

---

1   https://www.latimes.com/archives/la-xpm-1999-oct-01-mn-17288-story.html

adding damage to the institution's credibility and reputation, which could be the costliest loss of all.

Establishing a communication protocol:

Each project must have an established, well-structured communication protocol to be a "Road map" to follow during the lifetime of the project. Each project phase has its own communication plan and agent (s). In general, the plan should include:

1. **Import data plan (from previous phases):** Information necessary to carry the current project phase and allow a smooth, accurate transition and information transfer or exchange.

2. **Communication agent (s):** Individual or group, responsible for maintaining the communication flow open and running, but also a "Spokesperson" for the current phase. These agents are responsible of updating the stakeholders on the current project status and serve as liaisons or facilitators within the project team. Some of these potential project spokespersons are:
   - Project Manager.
   - EOR (Engineer Of Record).
   - Contractors.
   - Legal Rep.
   - Commissioning Manager.
   - Public Relation.

There are two ways of communication: written and verbal. Each one of them has its own benefits, and helps keep a positive communication flow for the project benefits. Let's briefly explore the attributes that can potentially be applied by the PMs:

1. **Written and Verbal Types:** Reporting, documenting, and storing reports are part of the project background and history, thus, determining reporting types, complexities, frequencies, audiences, confidentiality levels, and objectives must be well defined and communicated. Below are some typical (common) document checklist that a project manager can use to check and track (communicate) project progress:

    i. **Reporting documents and frequencies:** Via emails, shared drives, one notes, project management applications, or other means. The target groups vary depending on the industry, company's communication setup, and other applications. The list is mostly for guidelines and suggestions.

        a. Emails (on an as-needed basis) *.
        b. Daily. (Daily construction progress preferably with pictures)
        c. Weekly. (limited to team members and direct stakeholders).
        d. Monthly. (Typically, upper management)
        e. Quarterly. (Typically, upper management and executives).
        f. Yearly. (All stakeholders)
        g. Final (Finance and main stakeholders).

  ii. **Reporting document types:**

    a. Scope update (Engineering, Construction, Procurement, Permits, etc.).

    b. Financial update.

    c. Schedule update.

    d. Construction update.

    e. Safety Audit.

    f. A combination of all of the above.

  iii. **Stakeholders (Audience):**

    a. Executive.

    b. Project team.

    c. Local, State, and Federal agencies.

    d. Media.

    e. Residents.

2. **Macro vs Micro (When? How?):**

The answer is: **"Macro Manage Globally and Micro Manage Locally"** F. B.

A lot of supervisors, and project managers, unfortunately misuse both of these styles, by having an unbalanced communication style due to:

  i. Lack of experience.

  ii. Fear of the unknown and lack of knowledge.

  iii. The desire to assert authority rather than focusing on the project objective.

  iv. Overconfident or the opposite.

  v. One-way communication or "Giving orders" rather than sharing opinions.

    vi.   Fear of delegation.

   vii.   Lack of focus.

 viii.   Incompatibility with some team members.

    ix.   Lack of trust.

     x.   Overprioritizing: "Too many priorities, leads to no priority" F. Belguet.

    xi.   Overlooking issues.

   xii.   Focus on the shell of the project rather than the heart.

 xiii.   Lack of planning or anticipating.

 xiv.   "Listening" but not "Hearing".

   xv.   Poor choices in personal, strategy, and approach(es).

Macro Managing (Globally MMG):

One of the project manager's main duties is to create an "Autopilot" system that allows the project team members to focus on their duties in an "Automatic" way without interference or a lot of it. This approach will allow efficient and swift progress while freeing the project manager and executive to focus on the project strategy and the "Big picture," but also track any local deficiency due to specific circumstances and fix it before it negatively impacts the schedule, and, or the budget. Which leads to our next approach:

Micro Managing (Locally MML):

Before we detail this style, it is very important to emphasize that micro managing could be overburdened, tiresome, and overwhelming and could potentially have a negative impact on the personnel and their productivity. So, like any chemical reaction or medicine, a project manager has to be very careful about giving the right dosage of micromanaging team members based on:

i.   The severity of the issue.
ii.   The urgency.
iii. The capacity of the affected team member or sub-team.
iv.   The complexity of the specific task.
v.   Allocated support.
vi.   The effort and energy required.

The most important aspect of the micro-managing method is to not "Suck" all the air or overdramatize the situation. As such, it needs to be used in a particular way for the following:

i.   **Locally:** Spatiotemporal limitations. i.e., Limited time and space.
ii.   **On an "As needed" basis:** Avoid it as much as you can.
iii. **"Flag it" rather than micromanage it if possible:** Use the communication tools to express your concern (s) and give time for adjustment or resolution prior to micro-managing it.
iv.   **To address a "Specific"** task (s), group (s), or situation (s).

3. Records Keeping (History):

Like any living entity, a project has a beginning, a life, and an ending. However, the life cycle of a project goes on as long as its usage for what it was destined to accomplish. Thus, its "History" must be kept not just live, but updated during the life cycle. The history of the project starts with its inception, translated into record keeping and beyond.

Rather than giving a lengthy checklist on what is most likely to be tracked (or recorded), I've preferred to develop record blocks (RB) for project phases or at least give the project manager an organized way of selecting, modifying, and tracking project records as it is applicable to each specific field of expertise. The most important aspect is to develop project record blocks (RB) as early as possible, preferably at the beginning, so no record will be lost.

It is worth noting that these records are living, dynamic documents, subjected to updates under the following - but limited to – circumstances:

    i.   Technological updates.
   ii.   Safety enhancement.
  iii.   Security (Cyber or Field) enhancement.
  iv.   Intersecting other future projects.
    v.   Performance enhancement.
  vi.   Goals changes.

Below record blocks (RB) are typical templates (checklist) as a guideline to project record keeping. Naturally, for the application and field specifics, thus subjected to customization based on project needs:

Project Initiation	Project Scope Development	Finance
• Business Development Goals. • Feasibility studies.	• Control Narrative. • Engineering. • Geological and Environmental tests. • Location (s) confirmation. • Drawings: 1. IFRs. 2.IFCs. 3.As-built.	• Budget. • Forecast. • Procurement. • Proposals. • Contracts. • Change Orders. • Invoices.

Permits	Material & Equipment	Operation
• Local. • State. • Federal. • Eviromental. • Security. • Operation. • Legal.	• Bill of Material BOL. • Material references (Suppliers, heat#, etc). • Equipment info. (Suppliers, Warrenties, etc.). • Storage, Hauling, handling and freight tracking and ligistics. • Receipts. • Inventory & Parts lists.	• Manauals. • Personal trainingg. • O&M (Operation and Maintenence) Manuals. • Equipment tagging list. • Inventory & Parts tracking uploads tools (softwares & applications).

Communication:
- RFI.
- Submittals.
- Action Items list.
- Meeting minutes.
- Meetings audio.
- E-mails.
- Pictures.
- Videos.
- PR.

Schedule
- Initial Schedule.
- Baseline Schedule.
- Periodic updates.
- Final update.
- Child schedule (s).

Commissioning
- Commissioning team.
- Commissioning task.
- Partial & final commissioning schedule.
- Commissioning checklist.
- Commissioning approval/signee team.
- Performance Acceptance Protocol PAP..
- Performance Acceptance Test PAT.

Safety
- Safety Plans.
- Safety Procedures.
- Safety Work Plans (Site specfics)
- Safety Audits.
- Tool box recods.
- Near misses.

QA/QC
- Applicable Procedures.
- Applicable Codes & Best Practices
- NDE Testing Procdures & Records.
- Operators records and licenses.
- Pictures.
- Videos

4. Cultural awareness:

One of the most sensitive and overlooked aspects of communication—especially in international or regional affairs—is cultural awareness. Any project manager should be aware of his/her surrounding culture and be acquainted with local cultural customs and practices.

The second overlooked aspect is the notion of "Cultural differences (or Perceptions)." These are alive and very dynamic. They are in constant motion and up to changes, everywhere and every time. "Culture" is defined, according to Merriam-Webster dictionary, as:

i.   The customary beliefs, social forms, and material traits of a racial, religious, or social group. Also: the characteristic features of everyday existence (such as diversions or a way of life) shared by people in a place or time.

ii. The set of shared attitudes, values, goals, and practices that characterize an institution or organization.

iii. The set of values, conventions, or social practices associated with a particular field, activity, or societal characteristic.

iv. The integrated pattern of human knowledge, belief, and behavior depends upon the capacity for learning and transmitting knowledge to succeeding generations.[2]

This means it is impacted by a lot of factors, some of these cultural differences are shared below for informational purposes:

i.   **Local & National Cultures:** Wherever you are in the south or north, east or west, a project manager's

---

2   Merriam-Webster: https://www.merriam-webster.com/dictionary/culture#:~:text=%3A%20the%20customary%20beliefs%2C%20social%20forms,in%20a%20place%20or%20time

way of communication must involve an awareness of the region's specificity and customs in order the get the message through in the most effective way possible.

ii.   **National & International Cultures:** Thanksgiving and Christmas are "The holidays" in the US, and both in the Fall/Winter season, thus: when scheduling or budgeting for a project, a good PM should be aware of potential drops in efficiency and increase in overtime due to the holidays. However, if your project is in the Middle East or some Asian country, this season is actually the "Best" season to do work due to religious differences and climate conditions.

iii. **North & South Cultures:** The weather difference is significant enough that leads to completely an opposite approach! Winter in North America is Summer in Australia and vice versa!

iv.   **Generational and Technological Cultures:** Thanks to COVID-19 that helped boost the technological usage of "Hybrid" or sometimes 100% "Remote" forms of working, easily embraced and adopted by younger generations used to online communication tools (texting, emails, social media platforms, etc.). The way of managing a project has completely changed in shape or and tools, and a project manager must adapt and adopt the tools available in his/her hands and keep up with latest cultural change in communication.

Cultural awareness is certainly not limited to the above points, and ignoring them could lead to misunderstanding, miscommunication, and sometimes unfortunately to dire consequences regardless of the intentions (good or bad) or genuine behaviors. To better illustrate some of the cultural differences, below are a couple of examples of a "Clear" opposite point of view directly linked to local cultures:

i. **"La bise" vs "Ojigi":** There is no more striking difference in greeting than "La bise" in France or French culture and "Ojigi" in Japan. Both are ways of greeting in their respective countries or cultures. The French naturally give "La bise" cheek kisses to greet each other while in Japan they bow to each other "Ojigi" with no physical touch or contact, even with relatives. Each behavior is considered "Normal" and "Automatic" in their own respective cultures and environments, but try to switch them and you will realize that your behavior will be at least perceived as an "Odd" if not rejectable behavior.

ii. **Perfume test:** Keeping the same countries and looking at the same cultures, the perfume test is another striking example of cultural differences and perceptions. While having a good scent and wearing good perfume is considered as BC,

BG "Bon Chic Bon Genre", or "Good style, good class" and elegant in French culture, in Japan, it is completely the opposite! It is considered offensive and an invasion of privacy (or private space) to subject others to your own scent preferences. You must be as neutral as possible and limit the usage of your perfume to your own private life and space.

In such a diverse world, there are so many of these examples. You do not need to travel far to notice these kinds of nuances or big differences, that, sometimes, can "Make" or "Break" shakes your project communication flow and objectives.

Chapter 6:
Conclusion "Show time!"

Projection!

"Talent wins games, but teamwork and intelligence win championships".

Michael Jordan

Chapter 6:

Conclusion

This document is a compilation of very simple, comprehensive, rational, and proven guidelines that any new project manager can utilize to start his/her career with enough information and mechanisms to allow him/her to successfully navigate through any type of project, from software development to bridge construction.

A project manager is foremost a leader who builds a solid structure and can surround himself/herself with the right personnel and tools to achieve project goals safely and efficiently. But also, he/she prepares, predicts, prevents, and faces potential pitfalls as they surface along the road.

Project management is an exact science with tangibles and metrics, subjected to quantifiable goals and restrictions, but it is also an art as it is subjected to human and other natural or political and environmental factors that are unpredictable or unquantifiable, thus the challenge of creating a harmonious, balanced execution process that ultimately leads to a successful ending.

On the other hand, the reward is the rare satisfaction of raising a project from inception to conclusion and the joy of seeing the fruit of team effort and hard labor. It is an unmatched and unmeasurable feeling of satisfaction.

I would like to emphasize the importance of building a strong, harmonious team. A project manager cannot and should not know everything, but he/she should know how to build a team of professionals with enough expertise in the subject matter to achieve project goals.

I hope this simple guideline will help as many new project managers as possible and shed some light in the wilderness of their endeavors, toward fruitful careers and many successful projects.

"Management is the art of Planning, Predicting and Preventing"

F.M. Belguet